An epiphyte

Utricularia alpina, an unusual bladderwort

Epiphytes range in form from the delicate *Utricularia alpina*, a plant which adds to its diet by capturing microscopic animals in traps made from specially modified leaves, to robust strangling figs and bushy parasites. Most noticeable amongst this dependent flora are, in tropical America, members of the family Bromeliaceae. These bromeliads are nearly all 'tank' epiphytes, collecting rainwater in their overlapping leaves for their own use later on, and incidentally harbouring a substantial local fauna of snails, slugs, frogs, worms, and insects in the tanks. The pineapple, native of Brazil, is one of the few bromeliads that grow naturally on the ground. Ornamental variegated forms do not have edible fruits, but in Martinique the heads are cut and exported to France to be used for decoration.

Another epiphyte

Aechmea nudicaulis, common on trees in Trinidad

A bromeliad

An ornamental pineapple, *Ananas comosus,* var.

A miniature orchid (above)

Oncidium pusillum growing on the small twigs of trees, or even sometimes on leaves

An orchid (left)

The Cedros Bee orchid, *Oncidium lanceanum*

Very many kinds of showy epiphytic orchids may be found growing on the branches of trees in similar situations to those in which the 'wild pines' occur. These plants are not parasites, but simply use the branches for maintaining positions in the forest that are most favourable to them for obtaining light, water, and sufficient air.

The continuously warm humid conditions of tropical forests encourage the activity of fungi and other decomposing organisms, so that dead wood and leaves rot away quickly.

The floor of the forest is usually damp and shady—ideal conditions for ferns, mosses, and young climbers. *Selaginella* is related to the ferns. It grows in damp sheltered places where sunlight is diffused or broken up into flecks; light intensity is either very low or intermittent. Young climbers, like the aroid *Philodendron*, begin life near the ground and at first press their leaves tightly onto earth banks or the bark of tree trunks. As they climb up the trees, holding on by short roots, these plants change their form; the leaves become larger, spread out and arise further apart until, perhaps twenty or thirty feet (6–10 m) up, they begin to flower. Other aroids include native species of *Anthurium*, which behave in a similar way.

A fungus (left)

A Polypore on an old fence post

Selaginella (below)

Selaginella on a damp shady bank at Spring Hill, Trinidad

An Anthurium (above)

Anthurium andraeanum is known
in many horticultural varieties

According to conditions of light, shelter, and drainage, they live in the forest either as epiphytes or on the ground. *Anthurium andraeanum*, a native of Colombia, is often grown as a pot plant in the West Indies where it thrives in a free-draining soil in semi-shade.

Trinidad (left)

A young *Philodendron*, climbing
on a roadside cutting,
accompanied by *Trichomanes*
filmy ferns

Small clearings are formed when old trees die and fall. In these more open and better-lit places, as along the banks of streams, it is common to find tree ferns, the aroid *Spathiphyllum*, and clumps of *Heliconia*. Heliconias are large banana-like plants, much prized by tropical gardeners who have sufficient space and shelter to grow them. The Balisier, *Heliconia wagnerana*, is common in secondary forests in Trinidad.

Many possible combinations of ecological factors are responsible for the apparently unlimited number of forms of plant and animal life that are to be found in a tropical forest.

A Heliconia (above)

A variety of *Heliconia caribaea*.

The Maraval lily (left)

***Spathiphyllum cannifolium* is found along forest streams in Trinidad**

Another Heliconia

This species resembles the Balisier or Wild Plantain of Trinidad, which is very common in wet secondary situations

A Venezuelan species

Heliconia platystachya, **closely related to** *Heliconia marginata,* **a rare plant of southern Trinidad**

Savanna is the name given to those kinds of tropical vegetation, at low or moderately low elevation, which are dominated by grasses rather than by trees. In some savannas there are no trees at all; in others there may be some widely scattered or clumped small trees and shrubs. Savannas exist where the total annual rainfall is less than that needed for forest to develop; but they are also encouraged where the climate is strongly seasonal and there is a distinct period of drought, even though the total rainfall may be quite high. The effects of the dry season may be intensified by grazing and fire. Most broad-leaved trees are unable to survive long droughts, periodic burning, or the flooding which may occur on level land during the rainy season. Palms are often much more tolerant of these factors.

Savanna

The northern section of the Aripo Savanna, near Arima, Trinidad

A shrublet

Amasonia campestris, characteristic of the savannas of Trinidad and northern South America

Savanna ground orchid

Eulophia alta, widespread in tropical America and also in Africa

Some types of savanna-like vegetation have been created by clearing and subsequent grazing and burning. An example of this is the Grand Savanna on the west coast of Dominica. The St Joseph Savanna, near St Augustine in Trinidad, occupies a steep hillside where rapid drainage, heavy soil, and fire combine to provide conditions inimical to the growth of large trees. This savanna may have derived from a low woodland. Many scattered small trees and typical sun-loving herbs are present.

The White Aripo orchid

Otostylis brachystalix, also in South America

A very rare orchid in Trinidad

Habenaria pauciflora

The Aripo savannas of Trinidad are more natural. The land is almost perfectly flat, and at one time or another it may be flooded or completely dried out and baked hard. In these conditions, grasses, sedges and other small herbaceous plants predominate. Palms and shrubs form thickets along shallow watercourses.

The plants which grow naturally on the savannas are mostly quite different from those found in the forests. A few of them have showy flowers, and there are numerous ground orchids. These plants are perennial, and usually have regular seasonal flowering. They persist through dry periods by means of underground tubers or rhizomes.

Sundew

Drosera capillaris

In damp hollows on the open savannas one may find a species of sundew which is able to trap small insects by means of sticky glands on its leaves. The thickets are overtopped by Moriche palms, and the undergrowth is luxuriant and tangled. Scramblers and twiners with attractive flowers, such as the Savanna Flower, are quite common, but the Lady's Slipper orchid is a rare plant.

The Savanna Flower (above)

Mandevilla hirsuta is an attractive vine of the thickets and forest margins

Lady's Slipper (below)

Selenipedium palmifolium, the only Lady's Slipper orchid in the West Indies

Examination Vine (above)

Odontadenia grandiflora flowers in June at the end of the university year

A twiner (below)

Mesechites trifida grows in the thickets associated with the Aripo savannas

The driest islands of the southern Caribbean are the Netherlands Antilles, off the coast of Venezuela, and the Grenadines, lying between St Vincent and Grenada. These islands are dry enough for several kinds of cactus to be part of the natural vegetation. The larger species have either a columnar habit or the segmented mode of growth of the typical *Opuntia* or prickly pear. Other kinds scramble over bushes or climb on trees. Another form is the short globular Turk's Cap cactus, *Melocactus*. Eighty-seven species and several varieties of this cactus have been described from the island of Curaçao and its neighbours, Aruba and Bonaire.

The Turk's Caps are distinguished by the presence of an upper head covered with numerous needle-like spines and bearing the flowers and fruits. Cactus flowers are usually pollinated by flying insects. The inaccessibility of the small pink flowers of *Melocactus*, surrounded by their armour of sharp needles, is overcome by humming birds which hover unharmed above them while taking nectar. The fruits are also attractive to small birds which distribute the seeds.

The Turk's Cap cactus

Melocactus, a characteristic plant of the drier islands

For many people, the floristic essence of the humid tropics is enshrined in what are often loosely termed the flowering trees. The inference that trees with insignificant flowers, like oak, birch and beech, the native forest trees of northern latitudes, are somehow different and non-flowering is inescapable. The fact that many tropical trees also have small flowers is often overlooked by the less well-informed observer, who will equally willingly believe that such of them as bear useful or large fruits should be called fruit trees. These misconceptions ignore the simple fact that almost all the trees of the tropics produce flowers and fruits, however large or small these may be. Any other kind of tree would be a conifer; these are few and uncommon in the warmer parts of the world.

The Breadnut

Also known as Chataigne, this is the seed-producing variety of the Breadfruit, *Artocarpus altilis*

The Yellow Poui

Tabebuia rufescens, one of the several kinds of Yellow Poui

There is some excuse for error in places where, for example, coconut, breadfruit, *Casuarina* and Yellow Poui trees may all be seen growing together, and for believing that they do not seem to have much in common. They are all flowering plants and they all develop fruits. It is the acceptance of the facts of their fundamental similarity that makes the study of their differences so worthwhile.

The *Casuarina* bears a striking superficial resemblance to a conifer and is believed unquestioningly by many people to be one. There are several features that easily show it is not related to true pines at all, as the common name Whistling Pine would suggest. It is a native of Australia, widely planted in hot sandy places where other trees will not grow.

The breadfruit was brought to the West Indies from the Pacific at the end of the eighteenth century by Captain Bligh on his second voyage. As happens with many plants when they are introduced to new areas, varieties have arisen which have been selected by their growers. Quality of breadfruit fruits varies greatly but, because they do not seed, almost all of them must still be propagated from root suckers. The breadnut is well known in Trinidad where chataignes, the large chestnut-like seeds of this distinct variety, are an essential accompaniment to calaloo soup.

Coconuts, breadfruit, and *Casuarina*, although looking very different, have in common that their male and female reproductive organs are borne on the same tree, but in separate male and female flowers. This is not so with the Poui in which all the flowers are the same and each has both male and female parts. Pouis are all native in the American tropics, and many of them have local names. The Yellow Poui is called Apamata in Trinidad. Other species may have white, pink or purple flowers. Attractive to bees, they make a glorious if brief display, emphasised, at the end of the dry season, by the complete absence of leaves from the tree at flowering time.

Plants were used as sources of food, medicines, manufacturing materials, and ornaments long before a start was made on their scientific study. The result is that much of the lore of gardeners and farmers confuses both themselves and botanists, and much of what botanists say and write is meaningless to anyone except themselves.

Such topics as what is understood by a 'flowering tree' or a 'fruit tree' can be discussed indefinitely, especially in the Caribbean where horticulture and agriculture have evolved from different backgrounds and traditions. Arguments can range from misunderstood terminology and nomenclature to disagreement about the uses and effects of plant products.

Apart from genuine mistakes and areas of knowledge where research is still necessary, it is easy to disagree. For example, calling banana or pawpaw 'trees', refers to nothing more than the excessive size of these overgrown herbs. A banana has no wood; it is no more a tree than is a Canna Lily. In both, the 'stem' is made of overlapping leaf-stalks. A Canna is no more a lily than is a Water Lily. Water Lilies are not in any way related to lilies, nor are Water Hyacinths. However, Wind Flowers *are* lilies, but in the West Indies they are called Crocus, elsewhere they are called Rain Flowers, and in Barbados they are called Snowdrops!

Wind Flowers (left)

Two kinds of Wind Flower (*Zephyranthes*), sometimes incorrectly called Crocus

Barbados Cherry (above)

Also known as West Indian Cherry, *Malpighia punicifolia*

Confusion of names of plants, or of parts of plants, can be amusing, irritating, or even dangerous. Wild plants used in folk medicine are often euphemistically named when their therapeutic value is nil. Their properties may be only inferred because their names, such as Mint, Sage, or Balm, have been given to plants with superficial resemblance to European ones, but with totally different, maybe even poisonous, constituents. The survival of Amerindian names in Trinidad and Guyana reduces this risk because they are original words, often quite specific in their application. In Barbados and elsewhere, where the English language has been the only source of names, inappropriate transfers are quite common.

The West Indian or Barbados Cherry is indeed a West Indian plant, with several wild relatives in the natural woodlands and thickets of the islands. The fruit in some ways resembles a cherry, but botanically it is not at all related to the *Prunus* cherries of cooler regions. It is a pity that these local *Malpighia* fruits are not better known by their own name, because their culinary value, especially their very high vitamin C content, merits much wider local and international recognition.

Plant geographers have been unable to decide where Barbados Pride originally came from; certainly not from Barbados, but it is known by that name almost everywhere. The Barbados Gooseberry is a kind of climbing cactus and the Barbados Lilac is, needless to say, not a lilac. Perhaps the people of Barbados felt that they should create a flora to replace the native one that was cut down long ago.

Pride of Barbados

Caesalpinia pulcherrima

In the days before Columbus reached the West Indies and the era of European colonisation and influence began, there were very few plants in the natural floras of the islands from which human food could be obtained. The Arawaks and Caribs lived mainly by fishing; their cultivations were negligible. Almost all the familiar fruits and vegetables were introduced later from south or central America or from the Old World.

Similarly, most of the ornamental garden plants and decorative trees have been introduced from elsewhere. These have become so well established, in such numbers and diversity, that it is difficult now to appreciate that common shrubs and trees, like *Hibiscus, Bougainvillea, Cassia, Allamanda,* coffee, *Citrus*, Logwood, and Rain Tree, are not native.

The earliest introductions of food plants were, as one would expect, made from the neighbouring continent. It is likely that the Arawaks brought in corn and sweet potatoes. They probably also had coconut trees, although the true origin of this most important and useful palm is not known.

The *Capsicum* peppers are tropical American plants, from the wild species of which a great many different cultivated varieties have been derived. The Star Apple or Caimite, and the Naseberry or Sapodilla, are also American and were probably brought to the islands at an early date.

Perhaps the most important tree crop, originating in the tropical American forests, is cocoa, a plant with a history of domestication and use going back to the time long before Europeans knew that the New World existed. The Aztecs of Mexico thought the tree was of divine origin and Linnaeus took this up in naming it *Theobroma*, meaning 'the food of the gods'. The tree was introduced to Trinidad from Central America in 1625 by the Spaniards, who developed large plantations of it there. The first type that they had is called Criollo, a variety producing a very high quality pale-coloured bean with a non-astringent flavour. In about 1757, Dutch seamen brought, probably from Venezuela, another variety known as Amazonian Forastero or Amelonado. This is a high yielding more vigorous variety, but with beans of dark colour and astringent flavour.

After the two varieties had been grown together in Trinidad for some time, there arose by crossing a heterogeneous assemblage of cocoa plants known as Trinitario hybrids. These share features of both parental types in various combinations, of which the best selections have the vigour of Forastero associated with much of the quality of Criollo. The new varieties have become important for further breeding and improvement both in Trinidad and elsewhere.

The Wiri Wiri

Capsicum frutescens, one of the many varieties of perennial ornamental and culinary hot peppers

Cocoa hybrid

The external features of this pod are those of the Criollo variety, but the dark seed (inset) reveals a cross with Forastero

Besides the fruit, providing the commercial beans, the cocoa plant has many interesting botanical features. Among these, the beautiful red colour of the youngest leaves is a character shared by other tropical trees—for example, the highly ornamental *Bombax ellipticum*, another introduction from Central America—in which it serves to protect the immature leaves from harmful radiation in the bright sunlight.

Young leaves

The flush of copper-coloured new leaves on *Bombax ellipticum*, a common feature of tropical trees

The Cannonball tree (above)

The flower of *Couroupita guianensis*

Another feature of cocoa is the production of its flowers on the trunk of the tree, an adaptation of tropical plants with large heavy fruits which could not be carried on the finer branches. This is also a feature of the Cannonball Tree, an oddity introduced from Guyana into many botanical collections.

Nearer home, Trinidad is not only extremely fortunate in having a common wild shrub of exceptional intrinsic beauty, but has this in a very rare variety with double flowers. It is called *Warszewiczia* after the Polish botanist who first discovered it in Central America. In Trinidad, where it has been chosen as the national flower, it is better known as Chaconia. This name is commonly understood to commemorate the name of the last Spanish governor, Don José Maria Chacon, but there is a possible alternative origin. The wild Chaconia has numerous small flowers in clusters. Some of these develop an enlarged bright red sepal making the whole inflorescence appear as if decorated with small red ribbons or tags. In France and Spain in the eighteenth century, dancers had small brightly coloured silk or cotton ribbons sewn to their shirts, and in the nineteenth century similar flashes were used to decorate ladies' dresses. In Spain these were called 'chaconadas'. Perhaps French-speaking Trinidadians picked up the likeness and used this to refer to the plant which bore these ribbons. In this way they would have called it Chaconier in the same construction that the French use to refer to other important product-bearing plants as Cocotier, Bananier, Tamarinier or Callebassier.

The double variety of Chaconia, which has many more red sepals than the normal plant, was discovered in Trinidad in 1957 by David Auyong, and then brought into cultivation and propagated successfully by Royston Nichols.

The double Chaconia (left)

In cultivation in the botanical garden of the University of the West Indies, St Augustine, Trinidad

The nutmeg tree is special in that it produces two well-known spices—the large oily fragrant nutmeg seed, and the thin aromatic red strips of mace covering it. These spices were known in western Europe in the Middle Ages, but it was not until the Portuguese reached Banda in the Moluccas Islands in 1512 that Europeans discovered the source. Great conflicts between the Portuguese and Dutch arose from this discovery, and attempts were made to monopolise production by rigorous control of the growing and exportation of young trees.

A clandestine French expedition to the Moluccas in the eighteenth century obtained plants and shipped them to Mauritius and Cayenne. From these, introductions were made into Trinidad and St Vincent but none of them resulted in the establishment of rival industries. Later plantings in Grenada have, however, developed into a commercial enterprise of such success that that island stands to-day second only to Indonesia in the world production of these spices. Nutmeg is now the national emblem of independent Grenada which, because allspice, black pepper, ginger, Tonka bean, vanilla, cinnamon and clove are also grown there, is aptly named 'Spice island of the west'.

Other islands have their specialities. Dominica is renowned for the production of limes, and St Vincent for arrowroot.

Nutmeg

The ripe fruit of nutmeg, *Myristica fragrans*. The mace is the red aril covering the seed

There are a few beautiful private gardens and some historically interesting public ones in the West Indies. The main reason for creating them was to improve the availability of useful and ornamental cultivated plants in the region. The incentive to build up collections of plants has usually been provided by considerations of agricultural necessity or commercial opportunity, but, given wider choice, introductions have also been made of curious or rare plants with nothing more to commend them than their oddity or the challenge of bringing something new or strange into successful cultivation.

Mystical symbolism has always stimulated the imagination, and when this is seen in a startlingly attractive flower, the pleasure of growing the plant is greatly increased. *Passiflora* vines abound in the American tropics, and many of the more showy kinds are grown in gardens. When the Jesuit priests, who accompanied the Spanish Conquistadores to the New World, saw them they were reminded of the flower which St Francis of Assisi had seen growing on the cross in one of his visions. They referred to 'the flower of the five wounds' or Flos Passionis, recognising in the parts the symbols and instruments of the Passion.

The five sepals and five petals, usually all similar in shape and colour, were the ten faithful apostles; two were absent, Peter who deceived, and Judas who betrayed. The radiate corona was the crown of thorns, the five stamens were the five wounds, the ovary the hammer, the three styles the nails. Linnaeus later named the group *Passiflora*. Many species exist, including those which bear the delicious fruits known as Passion Fruits, Granadillas and Sweet Cups.

Botanically, the structure of these flowers is probably the result of some special method of pollination, as with the somewhat similar but unrelated Spider Plant, *Cleome speciosa*, which also has its anthers and stigma very prominently presented. This *Cleome* thrives in the garden at Spring Hill in Trinidad and other places at higher altitudes where the plants require cooler conditions. A rough comparison can be made between the growing conditions of a good European or North American summer and the conditions to be found generally at altitudes of two to three thousand feet (610—910 m) in these islands.

Passion Flower

Passiflora coccinea, **a native of Brazil**

Spider Plant

Cleome speciosa, Spider Plant, also known, in Barbados, as Ragged Robin

Plants which are grown out of doors in the temperate summer, will grow at these altitudes: it is too warm for most of them lower down. *Cleome, Canna,* and *Grevillea* are examples.

More difficult challenges are sometimes overcome successfully by tropical gardeners who are sufficiently determined. The writer's first sight of the Jade Vine was in the garden of the late Horace Gillette at St Augustine in Trinidad. This introduction from the Philippines was difficult to establish and propagate at first, but it is now becoming better known. Dr Gillette was one of very few who have mastered the cultivation of Sealing Wax palms in the West Indies. He had a large collection of orchids, one of which, the Malayan *Vanda teres*, escaped from his garden onto neighbouring trees.

A rampant grower

Vanda teres, a vine-type of orchid, with strong aerial roots which can attach themselves to rocks or trees

The Jade Vine

Strongylodon macrobotrys, a very attractive and vigorous climber, native of the Philippine Islands

Favourites with the horticulturists of the southern Caribbean are the many varieties of *Ixora* which have been developed there. All derived originally from wild plants brought from south-east Asia. *Ixora williamsii* was a name appearing in the horticultural literature of the last century for a plant grown by Mr B. S. Williams at the Victoria and Paradise Nursery, Upper Holloway, London. and recently identified with living examples from Trinidad. It has not so far been found in the wild state.

Ixoras

These are horticultural favourites in Trinidad and other islands

Lipstick (above)

A variety of *Ixora*

Ornamental cultivar (below)

Ixora williamsii, a plant of garden origin in Trinidad. The original showy Ixoras all came from south-east Asia and the Pacific area

Many tropical plants that have been transported to other countries where conditions are apparently similar, or even to hothouses in Europe, have proved easier to grow than are native plants in their own land. *Isertia parviflora* is a very handsome shrub resembling Ixora and common in the wild thickets of lowland Trinidad. It is a most attractive plant but extremely rare in cultivation anywhere. At the present time, West Indian gardeners, like Iris Bannochie in Barbados, have accepted this new challenge and are building up collections of decorative plants of local origin in their gardens.

Bois Caco

Isertia parviflora, a characteristic shrub of low-lying thickets and woodlands on the central plain of Trinidad

The Black Stick

Pachystachys coccinea, an attractive ornamental, but also a persistent, weed of cocoa farms in Trinidad

We tend to dislike weeds because they interfere with our horticulture or agriculture and often cost a great deal in time, effort, and money to control or remove. There are many kinds, and some of them demand special attention because of their peculiar characteristics, not always unpleasant. The selection of tropical weeds illustrated here all show unexpected and unusual features.

The Black Stick is a member of the large family Acanthaceae, a group of plants well represented in tropical gardens because of their showy flowers. This plant has escaped from gardens in Trinidad where it has become a persistent weed of cocoa farms.

Crepe Coq

Also known as Deer Meat,
Centropogon surinamensis.

The Crepe Coq or Deer Meat is a common native species from Guadeloupe southwards to Trinidad, and also on the continental mainland. It has similar weedy tendencies in damp shady places. It is an eye-catching plant but has to be treated with some respect as it belongs to the family Campanulaceae of which many members have poisonous substances associated with their milky sap.

Not all plants with latex are poisonous, but unless one knows which are and which are not, as with edible and poisonous fungi, it is best to be cautious in handling them and particularly to refrain from eating any part until quite sure that it is safe to do so.

Red Head

The flowers of the Red Head or Wild Ipecacuanha, *Asclepias curassavica*

The Wild Ipecacuanha or Red Head is a common weed with milky sap, known to be poisonous. In addition, it has several other points of interest. The extremely complex flowers, shaped like miniature crowns, are so formed that they can be pollinated only by certain kinds of insect. In moving over the flower in search of nectar, wasps carry the adhesive packets of pollen, which have become attached to their legs, from one plant to another. The device is similar to that used by orchids to ensure the same ends, but the two groups are not related. The seeds of this plant are windborne, being released one by one from the boat-shaped capsule where they lie in overlapping rows like roof-shingles. The botanical name of Red Head is *Asclepias curassavica*, the Asclepias from Curaçao. In the language of that island, which is a concoction of Dutch, Spanish and other elements, known as Papiemento, it is called Komchi cu skottel—Cup and Saucer.

From Curaçao

Wild Ipecacuanha or Red Head. Linnaeus called it *Asclepias curassavica* because it was first described from Curaçao

Crab's Eyes

Also called Jumbie Beads, ***Abrus precatorius,*** **a common weedy twiner in many tropical lands**

A plant which contains one of the most virulent of all plant poisons is the common weedy twiner known as Crab's Eyes or Jumbie Beads. This legume has very inconspicuous flowers, but the ripe pod opens and uncurls to expose the attractive two-coloured dangling seeds. These seeds are widely known. They have been strung as beads and were at one time used in India and Africa for weighing gold. They contain a dangerous protein poison, but they are also very hard and it is virtually impossible to crush them with the teeth. It is necessary to boil them before they can be pierced for stringing. This boiling destroys the poison, known as abrin, which must get into the bloodstream before it is harmful. Fatalities are consequently very rare, but nonetheless the seeds and necklaces made from them are banned from importation into some countries.

Castor Oil Plant

Ricinus communis has many varieties. This one is quite common along roadsides in the Windward Islands

The Castor Oil is another common plant with poisonous seeds, when untreated. There are many varieties of this shrubby weed—some of them worth cultivating as ornamentals; like the one on this page, which is frequently found along roadsides in Dominica, St Lucia, and other southern Caribbean islands.

Job's Tears (above)
Coix lacryma-jobi is a widespread grass of ditches and pond margins

Star Grass (right)
Rhynchospora nervosa, a common sedge

A widespread weedy grass, known as Job's Tears, has attractive shiny fruits that are often used as beads.

A lot of pasture, lawn and weedy fine-leaved plants are known as this or that kind of grass. Most of these are true grasses, but some of them belong to a different family, called sedges (Cyperaceae). Star Grass is a sedge found in open damp places on heavy soils. It is one of the few grass-like plants to be pollinated by insects, which are attracted by the showy bracts.

Many tropical weeds have a very broad distribution. This is not surprising considering that they are often vigorous plants with efficient means of dispersal and a wide tolerance of different environmental conditions. However, species native in India or Africa and other parts of the Old World did not arrive in the New World until after the voyages of Columbus.

Similarly, native tropical American species have been taken to the Old World only during the last four hundred or so years. The interchange of plants across the Atlantic Ocean continues in a mostly accidental and haphazard way. *Coccinia grandis*, originally from tropical Asia, has been established in Puerto Rico and Barbados for several years, but its advent to Trinidad was first recorded only as recently as 1973. In the same gourd, cucumber, melon and marrow family, Cucurbitaceae, is also placed one of the best known of all tropical weedy plants. This is the Cerasee or Carilla; the leaves are used universally in native medicine, especially to treat skin ailments and as an ingredient in tonic preparations. The fruit is used to make a pungent preserve. The fruit of a closely related cultivated variant is a popular vegetable.

From Sri Lanka (below)

An hitherto unrecorded weed from Trinidad, *Coccinea grandis*

Carilla (right)

Carilla or Cerasee, *Momordica charantia*, a common tropical vine

Index